TABLE OF CONTENTS

Nature's Balancing Act 4
Tropical Ecosystems 12
Ocean Ecosystems. 18
Tundra and Ice 26
Temperate Forests. 32
Deserts 38
Glossary 46
Index 47
Show What You Know 47
Further Reading 47
About the Author. 48

CHAPTER ONE

NATURE'S BALANCING ACT

A balanced **ecosystem** is like a working machine. Only instead of gears and rods, it's made up of all the living and nonliving things in an area.

Humans disrupt ecosystems in many ways, such as by building highways, clear cutting forests, and making pollution.

Taking EARTH'S Temperature

Ecosystem Effect

Jodie Mangor

Rourke Educational Media

rourkeeducationalmedia.com

Before & After Reading Activities

Before Reading:

Building Academic Vocabulary and Background Knowledge

Before reading a book, it is important to tap into what your child or students already know about the topic. This will help them develop their vocabulary, increase their reading comprehension, and make connections across the curriculum.

1. Look at the cover of the book. What will this book be about?
2. What do you already know about the topic?
3. Let's study the Table of Contents. What will you learn about in the book's chapters?
4. What would you like to learn about this topic? Do you think you might learn about it from this book? Why or why not?
5. Use a reading journal to write about your knowledge of this topic. Record what you already know about the topic and what you hope to learn about the topic.
6. Read the book.
7. In your reading journal, record what you learned about the topic and your response to the book.
8. After reading the book complete the activities below.

Content Area Vocabulary

Read the list. What do these words mean?

ecosystem

eroding

ewes

habitats

incubation

migration

permafrost

plankton

replensih

tropics

After Reading:

Comprehension and Extension Activity

After reading the book, work on the following questions with your child or students in order to check their level of reading comprehension and content mastery.

1. What are the three ways species can respond to climate change? (Summarize)
2. If the Earth's climate continues to change, what do you think the Earth's ecosystems will look like 50 years from now? (Infer)
3. How will rising temperatures affect desert ecosystems? (Asking Questions)
4. What types of ecosystems have you seen firsthand? (Text to Self Connection)
5. What are two qualities found in species most likely to adapt well to climate change? (Asking Questions)

Extension Activity

Choose an animal you are interested in. Find out what kind of ecosystem it lives in. Then research how climate change could affect this animal and its habitat. Make a food web diagram to show what you've learned.

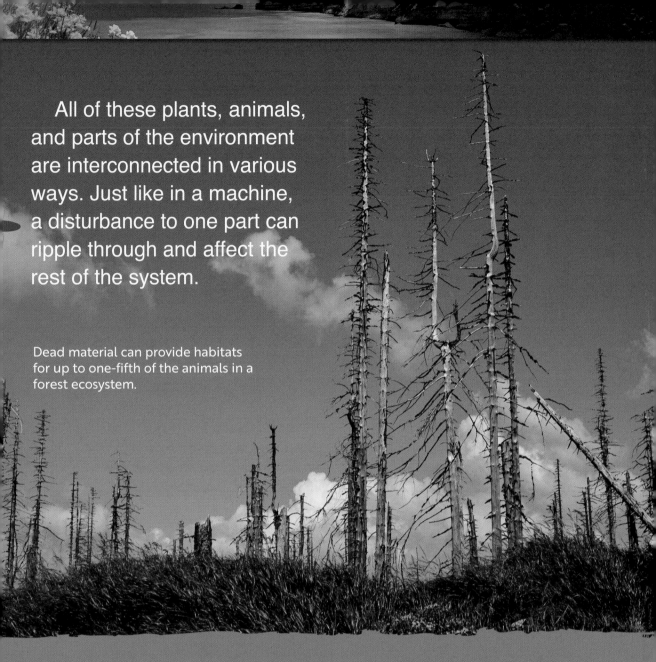

All of these plants, animals, and parts of the environment are interconnected in various ways. Just like in a machine, a disturbance to one part can ripple through and affect the rest of the system.

Dead material can provide habitats for up to one-fifth of the animals in a forest ecosystem.

One-third of woodland birds make their nests in dead trees.

If humans change a forest ecosystem by clearing away standing dead trees, the nuthatches and other insect-eating birds that nest in those trees will have no place to raise their young. Without birds to eat them, local insect populations will increase. This could spark even more changes to the ecosystem.

Snowshoe hares have large hind feet to help them travel on top of deep snow.

Climate is an important part of every ecosystem. Plants and animals that live in a particular climate have evolved to be suited for those conditions.

Climate and weather both describe local conditions, such as temperature, rainfall, cloudiness, and wind, of an area but over different time periods. Weather is what happens in a short period of minutes, days, or months. Climate is an average of weather conditions over a long period of time, such as many years.

Even small changes to the climate can have a big impact.

The amount and type of food that is available can change, causing species to shift where they live. New patterns of disease can arise. Seasonal behaviors such as breeding, **migration**, and hibernation can be affected. There might also be changes in growth and development, especially for plants and insects.

Most butterflies are very sensitive to weather and climate.

According to scientists, Earth's climate is changing more rapidly than ever before. Average temperatures are rising, precipitation patterns are changing, and there are more extreme weather events than before.

Spring is Arriving Earlier in the U.S.

There is a lot of scientific evidence that climate change is happening on a global scale. Climate change is already affecting ecosystems in measurable ways that vary from place to place. The cycling of water and nutrients is changing, affecting plants, animals, and insects. Overall temperatures are increasing. Conditions that lead to fire are more common than 50 years ago.

Some forests are becoming drier, making it easier for fires to start and spread.

There are three ways species can respond to climate change:

1. Adapt to the new environmental conditions.
2. Move to a more favorable climate.
3. Die. Some species will go extinct.

Climate change is altering ecosystems throughout the world.

Many plants and animals are migrating to higher elevations, where it is cooler. But species at the highest elevations don't have any place left to go to avoid rising temperatures. Island dwellers also don't have the option of going somewhere else.

There are still a lot of questions about how ecosystems will change, and how fast. The species that are adapting well to climate change can tolerate a range of climates. They tend to reproduce quickly and have genetic diversity. They can move to new **habitats** and compete with the species that are already there.

Jellyfish can live in a range of temperatures and are adapting well to climate change.

Genetic diversity has to do with the variety of genes in a species. Genes control inherited traits and behaviors. The more genetic diversity a species has, the more likely it is that some individuals in that species will have genes that can help them adapt to changes in the environment.

On the other hand, the species that are struggling tend to be more specialized, with narrow climate needs. Their populations lack genetic diversity. They may also face other pressures, such as being hunted, losing habitat due to human activity, and dwindling numbers.

Species in Australia and South America will be hit much harder than species in places with cooler climates, like North America and Europe.

Cheetahs have very little genetic diversity, increasing their risk of extinction.

CHAPTER TWO

TROPICAL ECOSYSTEMS

Tropical ecosystems are the most diverse on Earth. They support a rich variety of plants and animals.

Rainforests help stabilize the world's climate by absorbing the greenhouse gas carbon dioxide from the atmosphere. But large areas of rainforest are being chopped down and burned to make room for more agriculture. A lot of carbon dioxide is released in the process, which contributes to rising temperatures.

red-eyed tree frog

Tropical rainforests cover 6 percent of the Earth's surface and supply 40 percent of its oxygen.

rainbow lorikeet

Species in these ecosystems are already living at upper temperature limits. They are not used to big swings in temperature and may not be able to adapt to more heat. Scientists predict that by 2085, a third of all rainforest mammals, birds, reptiles, and frogs could become critically endangered or extinct.

Many rainforest species with specialized diets are at risk. This includes primates.

Because jaguars have broad diets, climate change isn't their most serious threat.

For now, climate change is helping some forest plants grow. Carbon dioxide gas is at atmospheric levels that are 40 percent higher than 100 years ago. Because plants use carbon dioxide to make food, some benefit from higher levels, up to a point.

Orangutans of the Southeast Asian islands of Borneo and Sumatra eat a lot of fruit. But as the rainforest becomes drier, its plants are producing less fruit. Forest fires are also becoming more common. These changes are forcing orangutans to move into new, less stable habitats.

Climate change is leading to food shortages for orangutans.

Of the 10,000 bird species in the world, 87 percent spend at least some time in the **tropics**. But increasing weather extremes are wrecking tropical habitats. When one species moves, other species in the ecosystem have to adjust or move themselves—if they can.

About 40 percent of the wild plants and animals being studied are moving to stay within a climate range they can tolerate. Some species can't move fast enough to get away from warming temperatures—or have no place cooler left to go.

Some birds are shifting their migration patterns.

In Costa Rica, keel-billed toucans and other lowland species have been moving to higher elevations to avoid rising temperatures. The toucans have to compete with another bird of paradise, the resplendent quetzal, for nesting sites in tree cavities. The toucans eat quetzal eggs and nestlings, further upsetting the balance.

Quetzals must learn how to protect their nests from toucans.

toucan

CHAPTER THREE
OCEAN ECOSYSTEMS

Oceans cover two-thirds of the planet's surface. They are home to many unique ecosystems and species. Ocean waters have absorbed more than 90 percent of the heat from climate change. As a result, ice is melting and ocean levels are rising. Fish reproduction, growth, development, and even survival are affected by higher temperatures.

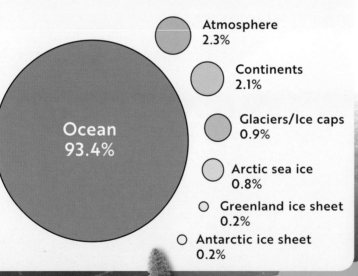

Earth's Surface

- Ocean 93.4%
- Atmosphere 2.3%
- Continents 2.1%
- Glaciers/Ice caps 0.9%
- Arctic sea ice 0.8%
- Greenland ice sheet 0.2%
- Antarctic ice sheet 0.2%

Right now, starfish are benefiting from climate change. As temperatures and carbon dioxide levels go up, they eat more and grow faster.

Wind patterns and ocean currents are also changing due to climate change. This affects species that follow currents when they migrate.

Whales, sharks, and sea turtles follow ocean currents to places where they feed.

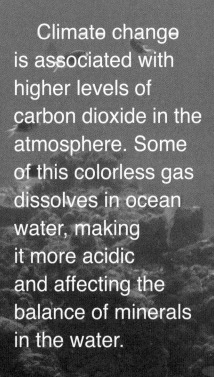

Climate change is associated with higher levels of carbon dioxide in the atmosphere. Some of this colorless gas dissolves in ocean water, making it more acidic and affecting the balance of minerals in the water.

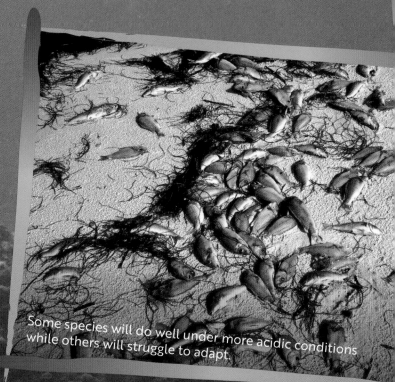

Some species will do well under more acidic conditions while others will struggle to adapt.

Higher levels of acid are a threat to coral reefs.

Lobster shell disease may spread with climate change.

Many shell-forming species such as oysters and clams are finding it difficult to build their shells. Corals and some **plankton** are also affected. Plankton are the base of food chains for many other ocean animals and fish.

As the ocean becomes more acidic, crab and lobster shells are getting harder.

Around the globe, fish are struggling with steadily rising ocean temperatures.

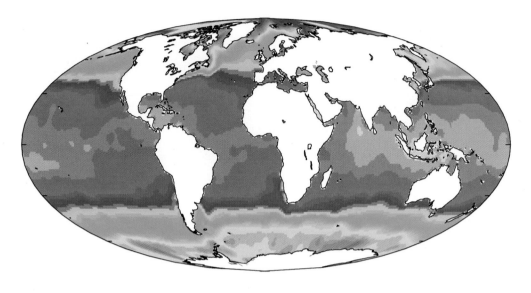

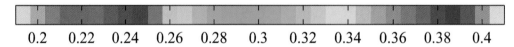

Sea−surface oxygen [mol O_2 m^{-3}]

0.2 0.22 0.24 0.26 0.28 0.3 0.32 0.34 0.36 0.38 0.4

Scientists predict many species of fish, including tuna, haddock, and cod, will shrink in size by as much as 30 percent due to warming temperatures and loss of oxygen in the sea.

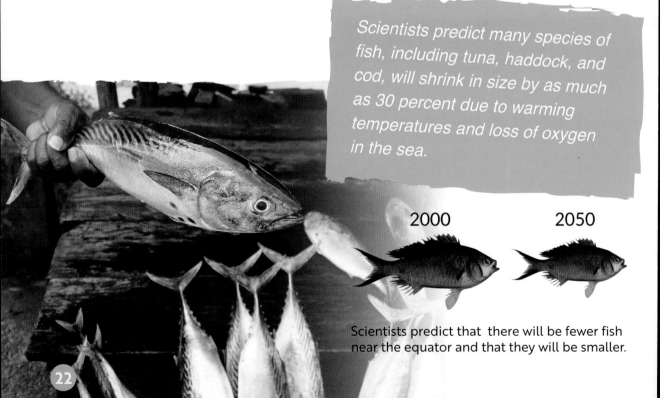

2000   2050

Scientists predict that there will be fewer fish near the equator and that they will be smaller.

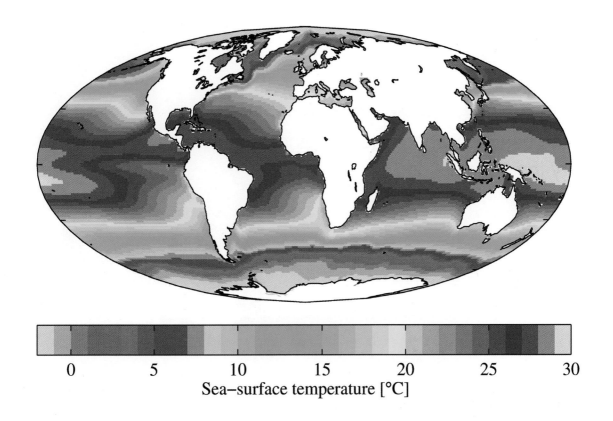

Tropical ocean fish are most at risk, since many already live near their upper temperature limit. Some ocean species are migrating closer to the North and South Poles to find cooler waters. Because everything is interconnected, this will impact populations that already live in these areas.

Some orcas are expanding their range to find enough prey to eat.

Sea turtles are among the many sea animals already affected by climate change. Rising seas and stormy weather are **eroding** many of the beaches where they lay and bury their eggs.

sea turtle eggs

If the sand gets hot enough, it can cause complete nest failure.

Also, the temperature of the sand is getting hotter. For sea turtles, the **incubation** temperature of an egg determines the sex of the hatchlings. Hotter sand means more females. In the short run, this will increase turtle populations. But if the number of females climbs too high compared to the number of males, the species could go extinct.

CHAPTER FOUR

TUNDRA AND ICE

So far, the Earth's coldest places have experienced the most global warming. This includes the Arctic, Antarctica, Alaska, parts of Canada, Scandinavia, and Russia. Arctic temperatures are rising twice as fast as anywhere else. Arctic ecosystems are fragile. When one part is disturbed, the ripple effect is swift.

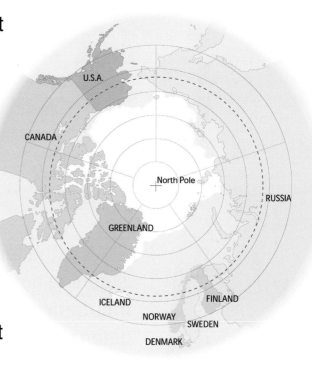

Huge amounts of ice have melted in the past 20 to 30 years. Ice shelves dating back tens of thousands of years have collapsed. Not only has this caused sea levels to rise, it is also a loss of essential habitat for many animals, including seals, polar bears, and penguins.

Sea levels are expected to rise between 7 and 23 inches (18 and 58 centimeters) by 2100.

Some species, such as whales, are benefiting from melting ice. Less ice means that more sunlight can reach the phytoplankton growing in the top layer of the ocean. This in turn means more food for whales. Less ice also makes it easier for whales from different areas to meet up with each other.

Polar bears are being hit hard by climate change and rapid loss of sea ice. Ice is an essential part of their habitat. They need it to hunt for seals and fish, find a mate, and raise cubs.

Increasingly, polar bears have to swim long distances, sometimes for days in a row, to find stable sea ice. This is especially difficult for cubs. As more ice melts, the polar bears have less chance of surviving.

Polar bears eat seals, which eat Arctic cod, which eat zooplankton, which eat ice algae. But as sea ice melts, the amount of ice algae is decreasing. The entire food chain will be affected by this change.

ACTIVITY

WHEN DOES MELTING ICE CAUSE SEA LEVELS TO RISE?

Polar ice can be found on land and in the ocean. How do these two types of ice affect sea levels when they melt?

YOU'LL NEED:

- Two identical clear containers that can hold about two cups (475 milliliters) of water
- Two lumps of modeling clay, each about the size of a fist
- Measuring cup
- Ruler
- Water
- 6 ice cubes
- Tape or marker

1. Put a lump of modeling clay into each container. Form each piece into a land mass that is pretty level on the top, and covers about half of the bottom of the container.

2. In the first container, put two ice cubes on top of the clay, lightly pressing them into place.

3. In the second container, put two ice cubes next to but not on the clay.

4. Pour water into each container, until the water levels are about equal but do not cover the top of the clay.

5. On the outside of each container, mark the water level with a piece of tape or a marker.

6. Allow the ice to melt, making sure that the melted water can drain off the clay.

7. When all of the ice has melted, check the water levels in each container again. Have they changed? What does this tell you about land ice and sea ice? Which can cause sea levels to rise when it melts?

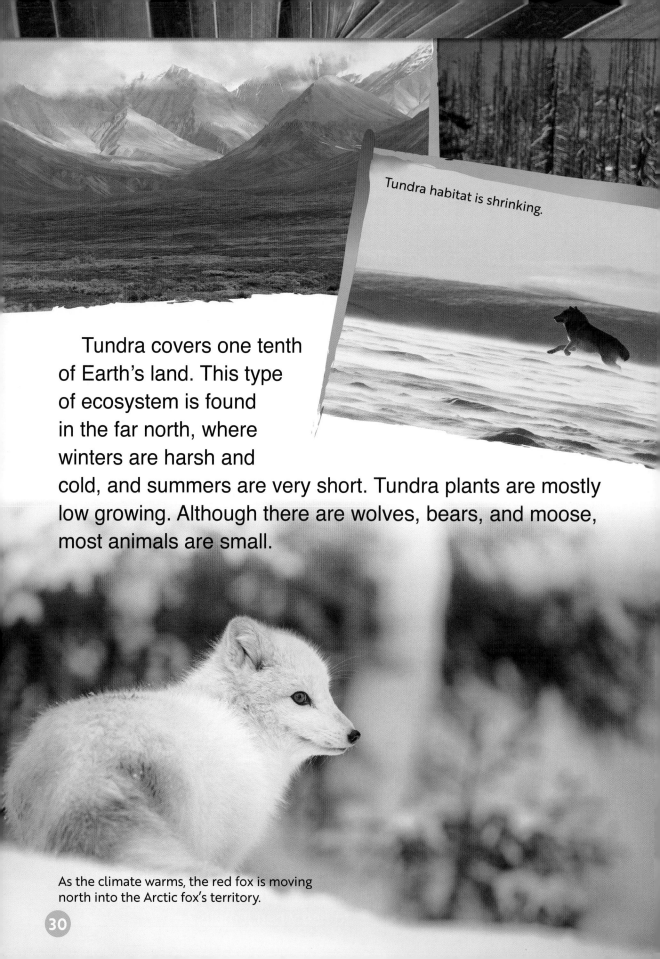

Tundra habitat is shrinking.

Tundra covers one tenth of Earth's land. This type of ecosystem is found in the far north, where winters are harsh and cold, and summers are very short. Tundra plants are mostly low growing. Although there are wolves, bears, and moose, most animals are small.

As the climate warms, the red fox is moving north into the Arctic fox's territory.

Warmer summers mean that the spruce bark beetle, which is only supposed to live for one summer, is now surviving for two. It is destroying many trees in Arctic forests.

When Permafrost Thaws

1. High levels of greenhouse gases in the atmosphere cause global temperatures to increase
2. Increasing temperatures cause permafrost to thaw
3. Thawing exposes previously frozen organic matter to decay
4. As organic matter decays it releases CO_2 (and sometimes methane) into the atmosphere
5. Atmospheric CO_2 and methane levels increase

Much of the soil stays frozen year round, but this **permafrost** has started to melt. As the frozen soil thaws, it releases greenhouse gases like carbon dioxide and methane that contribute to climate change.

CHAPTER FIVE

TEMPERATE FORESTS

Temperate forests are found where there are four seasons. Humans have cut down many of these forests and climate change is affecting those that remain.

70,000 square miles (181,300 square kilometers) of Rocky Mountain conifers died when weakened trees were infected with insects.

In Europe, wild boar populations are soaring. Climate change is causing trees to produce more of the acorns and chestnuts wild boars eat. Warmer winters are helping piglets born near the start of winter to survive.

Warmer winters mean there is less snow to **replenish** water supplies. Longer, hotter summers can lead to drought and fires. As forest ecosystems become more stressed, the trees are weakened. Tree-killing insects are thriving and spreading under these new conditions.

Aquatic insects are an important part of the food web.

Amphibians need moist habitats and are sensitive to changes in temperature and precipitation. Many breed in ponds and seasonal pools. As drought becomes more common, these water sources will dry up faster and be available for a shorter time. This means there will be fewer aquatic larvae for amphibians to eat.

Wood frogs breed and lay their eggs in seasonal pools.

For many species, death rates will be higher. On the other hand, warmer winters may help some species survive.

Insects are an important part of ecosystems. They are a source of food for many animals. For example, small bats of the Northeast eat flying insects. Many of these insects spend part of their life cycle in water. Less water due to drought means fewer insects, which means less food for bats. If these bats aren't able to store enough energy to last through hibernation, they may not survive.

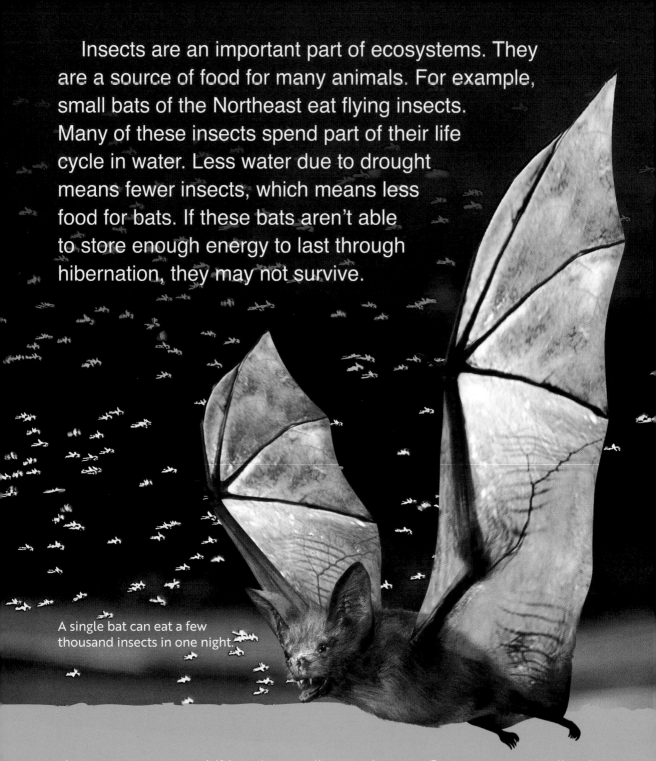

A single bat can eat a few thousand insects in one night.

Insect ranges are shifting due to climate change. Some are expanding into new areas. Studies show many non-migratory butterflies live farther north than they did 50 years ago.

ACTIVITY

MAKE A FOOD WEB BASED ON A LOCAL ECOSYSTEM

A food web diagram shows the relationships between organisms in an ecosystem. To make one, write down at least ten animals, plants, and other organisms that live in your neighborhood on slips of paper. Now divide them into the following groups:

- **Producers:** Plants are producers because they are able to capture light energy from the sun to produce food for themselves and other organisms.

- **Primary Consumers:** These are animals that eat only plants.

- **Secondary Consumers:** These are animals that eat other animals.

- **Tertiary Consumers:** These are animals that eat secondary consumers.

- **Decomposers:** This group includes organisms such as bacteria, fungi, and worms that can break down organic matter. This allows plants to use the nutrients that are released back into the food chain.

Glue each slip on a separate piece of paper and draw arrows to show the flow of energy (what eats what) in your food web. The arrows should point from the sun to producers, from producers to consumers, and from consumers to consumers.

CHAPTER SIX

DESERTS

Desert areas make up almost a quarter of the Earth's surface. Climate change is expected to increase the number of deserts. It will also impact the high level of animal diversity found in this type of ecosystem.

Most deserts get less than ten inches (25.4 centimeters) of precipitation each year.

Small changes in temperature or precipitation could have a huge effect on plants and animals living in the desert. Rainfall is expected to decrease by as much as 20 percent. This will lead to more drought, dust storms, and wildfires. Unique habitats and rare species could disappear.

Desert spiny lizards need many small patches of shade to thrive.

The giant ground gecko of South Africa.

Desert animals are often very small to help them regulate body temperature. Most are active at night when it is cooler. Water is scarce, so shrubs and plants have evolved to minimize water loss. Cacti can absorb and store water. This allows them to survive long periods of drought.

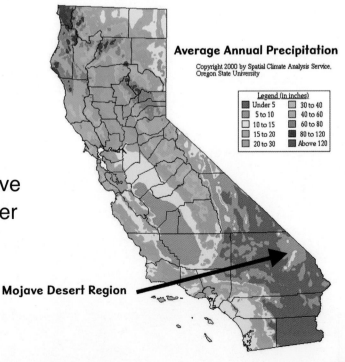

Average Annual Precipitation

Copyright 2000 by Spatial Climate Analysis Service, Oregon State University

Legend (in inches)
- Under 5
- 5 to 10
- 10 to 15
- 15 to 20
- 20 to 30
- 30 to 40
- 40 to 60
- 60 to 80
- 80 to 120
- Above 120

Mojave Desert Region

Climate change could wipe out one-fifth of all lizard species by 2080. Reptiles can't regulate their own body temperatures. Many species already live at the edge of their thermal limits.

When it heats up, lizards rest in the shade to cool down.

When temperatures are high, they need to find a cool place to rest. They can't spend as much time foraging for food or mating. As days grow hotter, some lizards may need to become nocturnal to survive.

Bearded dragon lizards that come from eggs incubated at higher temperatures are less intelligent.

Researchers predict that nearly half of the bird, mammal, and butterfly species in southwestern deserts will be replaced by other species by 2055.

Desert bighorn sheep are threatened by climate change. They have adapted to live in the dry, desert mountains of the southwestern United States. The steep, rocky slopes keep them safe from predators.

30 local populations of desert bighorn sheep in California have already gone extinct.

But rising temperatures and less precipitation is decreasing the quality and quantity of the vegetation they eat. It's also harder for them to find water. This is a problem for nursing **ewes** who need to drink more often to make milk for their young.

Citizen scientists collect many types of data to help study climate change.

Climate change is already affecting ecosystems around the globe. Scientists predict greatly reduced diversity in most parts of the world by the year 2080.

What is citizen science? It's when volunteers and scientists work together to gather data to answer real-world questions. This way, scientists are able to obtain more data from more places, over longer periods of time than if they were working alone. Volunteers get to learn about important topics in the natural world firsthand.

What Can We Do?

- Pay attention to what is happening in the ecosystems around you.

- Join a citizen science project.

- Do what you can to protect the environment, preserve natural resources and reduce greenhouse gases that contribute to climate change.

  - Use less heat and air conditioning.

  - Remember to turn off lights and other appliances when you're not using them.

  - Reduce the amount of stuff you buy and garbage you make.

  - Reuse materials instead of throwing them away.

  - Recycle what you can.

Glossary

ecosystem (EE-koh-sis-tuhm): a community of living things interacting with their environment

eroding (i-RODE-ing): gradually wearing away by water or wind

ewes (yooz): female sheep

habitats (HAB-i-tats): places where certain plants or animals are normally found

incubation (ING-kyuh-bay-shuhn): the act of keeping eggs warm before they hatch

migration (mye-GRAY-shuhn): movement of people or animals from one region or habitat to another

permafrost (pur-MUH-frawst): a permanently frozen layer below the Earth's surface

plankton (PLANGK-tuhn): tiny animals and plants that drift or float in oceans and lakes

replenish (ree-PLEN-ush): to make full or complete again

tropics (TRA-piks): the hot region of the Earth near the equator

Index

bird(s) 5, 13, 16, 17, 41
carbon dioxide 12, 14, 18, 20, 31
disease 7
drought 33, 34, 36, 38, 39
environment 5, 10, 45
fire(s) 8, 15, 33
forest(s) 5, 14, 15, 31, 32, 33
ice 18, 26, 27, 28, 29
insect(s) 5, 7, 8, 33, 36
lizard(s) 40, 41
polar bears 27, 28
precipitation 7, 34, 38, 43
primates 14
sheep 42, 43

Show What You Know

1. Name three ways climate change is altering ecosystems.
2. Compare the characteristics of species that are adapting well to climate change to those that aren't adapting well.
3. What are two ecosystems that have high animal diversity?
4. How are rising ocean temperatures affecting ocean ecosystems?
5. Give an example of how a change to one part of an ecosystem affects the rest of that system.

Further Reading

Collard, Sneed B. III, *Hopping Ahead of Climate Change: Snowshoe Hares, Science, and Survival*, Bucking Horse Books, 2016.

Coutts, Lyn, *Global Warming*, Barron's Educational Series, 2017.

Kurlansky, Mark, *World Without Fish*, Workman Publishing Company, 2014.

About the Author

Jodie Mangor writes magazine articles and books for children. She is also the author of audio tour scripts for high-profile museums and tourist destinations around the world. Many of these tours are for kids. She lives in Ithaca, New York, with her family.

© 2019 Rourke Educational Media

All rights reserved. No part of this book may be reproduced or utilized in any form or by any means, electronic or mechanical including photocopying, recording, or by any information storage and retrieval system without permission in writing from the publisher.

www.rourkeeducationalmedia.com

PHOTO CREDITS: istock.com, shutterstock.com, Cover: Leaf background © Twinster photo, Hare © Mark Medcalf, Ocean © Damsea; Pg4; Peter_Virag, Velvetfish. PG5: PETR BONEK , Creator: MattCuda. PG6: twildlife, impr2003, BenGoode. PG7: cicloco, USO, staphy. PG8: NeilLockhart, ArtMechanic, stsmhn. PG9: Renita Colaco, RPBMedia, Aunt_Spray. PG10: USO. PG11: By Eric Isselee PG12: Jag_cz, Ramdan_Nain, Dynamicfoto-PedroCampos PG13: gopfaster, PG14: Bossiema, USO, Photocech. PG15: USO, Brandon Silva, Lillian Tveit. PG16; tahir abbas, Lucas Pacheco. PG17: Mats Lindberg, robybenzi, GCastellon, Uwe-Bergwitz. PG18: leonello, mihtiander, OlgaLiss. PG19: Capt javed patel, Grafner, miblue5. PG20: Placebo365, EyeMark. PG21: Wrangel, comzeal, Artist: FishTales. Pg22; Plumbago, sal73it. PG23: Plumbago, kjekol. PG24: MartinPBGV, jarnogz, davidevison. PG25: danikancil, RedWhiteCreative, paulprescott72. PG26: FreeTransform, Kingrobby. PG27: JPL.gov, NASA, vladsilver. PG28: JohnPitcher, PeterZwitser, SeppFriedhuber. PG30; mlharing, Neil_Burton, troutnut. PG31: iChip, lantapix. PG32: Sean Pavone, Ohotnik, PhilAugustavo. PG33: KaraGrubis, Andy_Hu. PG34: woodygraphs, Ocs_12, JasonOndreicka. PG35: MikeLaptev. PG36: KajaNi, Geerati. PG38: Givaga, wrangel, bluebeat76. PG39: Oregon State University, Antonel, vaeenma, EcoPic. PG40: HuyThoai. PG41: Stacy Schenkel, CamiloTorres, tntphototravis. PG42: mikexavier. PG43: Jeff439, equigini. PG44: KeithSpaulding, EvgeniyShkolenko. PG45: Artranq, RTimages

Edited by: Keli Sipperley

Produced by Blue Door Education for Rourke Educational Media. Cover and Interior design by: Jennifer Dydyk

Ecosystem Effect / Jodie Mangor
(Taking Earth's Temperature)
 ISBN 978-1-64156-449-6 (hard cover)
 ISBN 978-1-64156-575-2 (soft cover)
 ISBN 978-1-64156-693-3 (e-Book)
Library of Congress Control Number: 2018930476

Rourke Educational Media

Printed in the United States of America, North Mankato, Minnesota